NUCLEAR ENERGY AND INFORMATION

RADIATION IN PERSPECTIVE

Applications, Risks and Protection

NUCLEAR ENERGY AGENCY
ORGANISATION FOR ECONOMIC CO-OPERATION AND DEVELOPMENT

ORGANISATION FOR ECONOMIC CO-OPERATION AND DEVELOPMENT

Pursuant to Article 1 of the Convention signed in Paris on 14th December 1960, and which came into force on 30th September 1961, the Organisation for Economic Co-operation and Development (OECD) shall promote policies designed:

— to achieve the highest sustainable economic growth and employment and a rising standard of living in Member countries, while maintaining financial stability, and thus to contribute to the development of the world economy;
— to contribute to sound economic expansion in Member as well as non-member countries in the process of economic development; and
— to contribute to the expansion of world trade on a multilateral, non-discriminatory basis in accordance with international obligations.

The original Member countries of the OECD are Austria, Belgium,Canada, Denmark, France, Germany, Greece, Iceland, Ireland, Italy, Luxembourg, the Netherlands, Norway, Portugal, Spain, Sweden, Switzerland, Turkey, the United Kingdom and the United States. The following countries became Members subsequently through accession at the dates indicated hereafter: Japan (28th April 1964), Finland (28th January 1969), Australia (7th June 1971), New Zealand (29th May 1973), Mexico (18th May 1994), the Czech Republic (21st December 1995), Hungary (7th May 1996), Poland (22nd November 1996) and the Republic of Korea (12th December 1996). The Commission of the European Communities takes part in the work of the OECD (Article 13 of the OECD Convention).

NUCLEAR ENERGY AGENCY

The OECD Nuclear Energy Agency (NEA) was established on 1st February 1958 under the name of the OEEC European Nuclear Energy Agency. It received its present designation on 20th April 1972, when Japan became its first non-European full Member. NEA membership today consists of all OECD Member countries, except New Zealand and Poland. The Commission of the European Communities takes part in the work of the Agency.

The primary objective of NEA is to promote co-operation among the governments of its participating countries in furthering the development of nuclear power as a safe, environmentally acceptable and economic energy source.

This is achieved by:

— *encouraging harmonization of national regulatory policies and practices, with particular reference to the safety of nuclear installations, protection of man against ionising radiation and preservation of the environment, radioactive waste management, and nuclear third party liability and insurance;*
— *assessing the contribution of nuclear power to the overall energy supply by keeping under review the technical and economic aspects of nuclear power growth and forecasting demand and supply for the different phases of the nuclear fuel cycle;*
— *developing exchanges of scientific and technical information particularly through participation in common services;*
— *setting up international research and development programmes and joint undertakings.*

In these and related tasks, NEA works in close collaboration with the International Atomic Energy Agency in Vienna, with which it has concluded a Co-operation Agreement, as well as with other international organisations in the nuclear field.

Publié en français sous le titre :
LE POINT SUR LES RAYONNEMENTS
Applications, risques et protection

Cover: Model of the repository for low- and medium-levels radioactive waste at Olkiluoto, Finland.
Credit: TVO, Finland.

$$= 7.43 \text{ mg/L} - 4.21 \text{ mg/L}) \times \frac{300 \text{ mL}}{10 \text{ mL}}$$

$$= 96.6 \rightarrow 97 \text{ mg/L}$$

5.

$$\text{Depletion} = \underset{\text{(Seeded Eff)}}{(DO_i - DO_f)} - \left[\underset{\text{(Seed Alone)}}{(DO_i - DO_f)} \times \frac{(\text{\% Seed in Effluent})}{(\text{\% Seed Alone})}\right]$$

$$= (7.29 \text{ mg/L} - 4.62 \text{ mg/L})$$

$$- (7.58 \text{ mg/L} - 4.89 \text{ mg/L}) \times \frac{(0.6\%)}{(4\%)}$$

$$= 2.27 \text{ mg/L}$$

$$\text{BOD} = \frac{\text{Depletion}}{\text{Sample Fraction}} = \frac{2.27 \text{ mg/L}}{0.18} = 12.6 \rightarrow 13 \text{ mg/L}$$

6.

$$\text{Depletion} = \underset{\text{(Seeded Eff)}}{(DO_i - DO_f)} - \left[\underset{\text{(Seed Alone)}}{(DO_i - DO_f)} \times \frac{(\text{\% Seed in Effluent})}{(\text{\% Seed Alone})}\right]$$

$$= (6.86 \text{ mg/L} - 3.95 \text{ mg/L})$$

$$- (7.44 \text{ mg/L} - 4.62 \text{ mg/l}) \times \frac{(0.5\%)}{(5\%)}$$

$$= 2.63 \text{ mg/L}$$

$$\text{BOD} = \frac{\text{Depletion}}{\text{Sample Fraction}} = \frac{2.63 \text{ mg/L}}{0.25} = 10.52 \rightarrow 11 \text{ mg/L}$$

---

7. DO_i of the seeded effluent sample

$$\text{Dilution water fraction} = 1 - (0.12 + 0.01)$$

$$= 1 - 0.13$$

$= 0.87$

Initial DO $= (8.1 \text{ mg/L} \times 0.87) + (5.2 \text{ mg/L} \times 0.12)$

$+ (2.0 \text{ mg/L} \times 0.01)$

$= 7.69 \text{ mg/L}$

DO_i of the seed alone sample

Dilution water fraction $= 1 - 0.06$

$= 0.94$

Initial DO $= (8.1 \text{ mg/L} \times 0.94) + (2.0 \text{ mg/L} \times 0.06)$

$= 7.73 \text{ mg/L}$

---

8. DO_i of the seeded effluent sample

Dilution water fraction $= 1 - (0.18 + 0.006)$

$= 1 - 0.186$

$= 0.814$

Initial DO $= (7.9 \text{ mg/L} \times 0.814) + (4.8 \text{ mg/L} \times 0.18)$

$+ (0.0 \text{ mg/L} = 0.006)$

$= 7.29 \text{ mg/L}$

DO_i of the seed alone sample

Dilution water fraction $= 1 - 0.04$

$= 0.96$

Initial DO $= (7.9 \text{ mg/L} \times 0.96) + (0.0 \times 0.04)$

$= 7.58 \text{ mg/ L}$

---

9. DO_i of the seeded effluent sample

$$\text{Dilution water fraction} = 1 - (0.18 + 0.01)$$

$$= 1 - 0.19$$

$$= 0.81$$

$$\text{Initial DO} = (8.1 \times 0.81) + (4.6 \times 0.18) + (1.5 \times 0.01)$$

$$= 7.40 \text{ mg/L}$$

DO_i of the seed alone sample

$$\text{Dilution water fraction} = 1 - 0.08$$

$$= 0.92$$

$$\text{Initial DO} = (8.1 \times 0.92) + (1.5 \times 0.08)$$

$$= 7.57 \text{ mg/L}$$

$$\text{Depletion} = \underset{\text{(Seeded Eff)}}{(DO_i - DO_f)} - \left[\underset{\text{(See Alone)}}{(DO_i - DO_f)} \times \frac{(\text{\% Seed in Effluent})}{(\text{\% Seed Alone})}\right]$$

$$= (7.40 \text{ mg/L} - 4.23 \text{ mg/L}) - (7.57 \text{ mg/L} - 4.92 \text{ mg/L}) \times \frac{(1\%)}{(8\%)}$$

$$= 2.84 \text{ mg/L}$$

$$\text{BOD} = \frac{\text{Depletion}}{\text{Sample Fraction}} = \frac{2.84 \text{ mg/L}}{0.18} = 15.78 = > 16 \text{ mg/L}$$

---

10. DO_i of the seeded effluent sample

$$\text{Dilution water fraction} = 1 - (0.12 + 0.004)$$

$$= 1 - 0.124$$

$$= 0.876$$

$$\text{Initial DO} = (7.96\ \text{mg/L} \times 0.876) + (2.43\ \text{mg/L} \times 0.12) + (0.0\ \text{mg/L} \times 0.004)$$

$$= 7.26\ \text{mg/L}$$

DO_i of the seed alone sample

$$\text{Dilution water fraction} = 1 - 0.05 = 0.95$$

$$\text{Initial DO} = (7.96\ \text{mg/L} \times 0.95) + (0.0\ \text{mg/L} \times 0.004)$$

$$= 7.56\ \text{mg/L}$$

$$\text{Depletion} = \underset{\text{(Seeded Eff)}}{(DO_i - DO_f)} - \left[\underset{\text{(Seed Alone)}}{(DO_i - DO_f)} \times \frac{(\%\ \text{Seed in Effluent})}{(\%\ \text{Seed Alone})}\right]$$

$$= (7.26\ \text{mg/L} - 3.98\ \text{mg/L}) - (7.56\ \text{mg/L} - 4.12\ \text{mg/L}) \times \frac{(0.4\%)}{(5\%)}$$

$$= 3.00\ \text{mg/L}$$

$$\text{BOD} = \frac{\text{Depletion}}{\text{Sample Fraction}} = \frac{3.00\ \text{mg/L}}{0.12} = 25\ \text{mg/L}$$

---

11. $$\text{Normality(1)} \times \text{Volume(1)} = \text{Normality(2)} \times \text{Volume(2)}$$

$$0.05\ \text{N} \times 2000\ \text{mL} = 10\ \text{N} \times \text{Volume(2)}$$

$$\frac{0.05\ \text{N}}{10\ \text{N}} \times 2000\ \text{mL} = \frac{10\ \text{N}}{10\ \text{N}} \times \text{Volume (2)}$$

$$\frac{0.05\ \text{N}}{10\ \text{n}} \times 2000\ \text{mL} = \text{Volume (2)}$$

$$= 10\ \text{mL}$$

12. $$\text{Normality}(1) \times \text{Volume}(1) = \text{Normality}(2) \times \text{Volume}(2)$$

$$5\text{ N} \times 10\text{ mL} = \text{Normality}(2) \times 8.4\text{ mL}$$

$$5\text{ N} \times \frac{10\text{ mL}}{8.4\text{ mL}} = \text{Normality}(2)$$

$$\frac{5\text{ N} \times 10\text{ mL}}{8.4\text{ mL}} = \text{Normality}(2)$$

$$= 5.95\text{ N}$$